YOUR KNOWLEDGE HAS VALUE

- We will publish your bachelor's and master's thesis, essays and papers

- Your own eBook and book - sold worldwide in all relevant shops

- Earn money with each sale

Upload your text at www.GRIN.com and publish for free

Bibliographic information published by the German National Library:

The German National Library lists this publication in the National Bibliography; detailed bibliographic data are available on the Internet at http://dnb.dnb.de .

Imprint:

Copyright © 2015 GRIN Verlag
Print and binding: Books on Demand GmbH, Norderstedt Germany
ISBN: 9783668814264

This book at GRIN:

https://www.grin.com/document/441711

Florante Jr Poso

Bouyancy. The Archimedes Principle

GRIN Verlag

Table of Contents

	Page
Introduction	3
Buoyant Force	3
Archimedes Principle	4
Buoyant Force Equation	5
Problem Solving	8
Summary	12
References	13

<u>**Introduction**</u>

The eBook discusses the Archimedes principle of buoyancy and the buoyancy equation in general. Application to the field of engineering was also expounded in order to show the relevance of the principle in the engineering context.

Sample problems are presented to understand fully the application of the buoyancy principle of Archimedes. Analysis of whether a certain object will float or sink are then explained based on the buoyancy equation. Therefore stability of objects can be analyzed by applying the mentioned principle.

The principle of buoyancy can be applied in floating objects such as ships and boats, submarines, hydrometer, balloons and airships and so many other real-life applications.

<u>**Buoyant Force**</u>

"A buoyant force is defined as an upward force (with respect to gravity) on a body that is totally or partially submerged in fluid, either a liquid or gas. Buoyant forces are caused by the hydrostatic pressure distribution."[1]

"When a solid object is wholly or partly immersed in a fluid, the fluid molecules are continually striking the submerged surface of the object. The forces due to these impacts can be combined into a single force, the buoyant force."[2]

"The buoyant force, which always opposes gravity, is nevertheless caused by gravity. Fluid pressure increases with depth because of the (gravitational) weight of the fluid above. This increasing pressure applies a

---

[1] Donal F. Elger, et. Al, <u>Engineering Fluid Mechanics</u>, 10th Edition, Wiley, USA, 2013, p 86

[2] <u>Bouyancy: Floating and Sinking,</u> Visual Physics, School of Physics, University of Sydney, Australia, *src.gov.jm/wp-content/uploads/2012/12/buoyancy.pdf*

force on a submerged object that increases with depth. The result is buoyancy."[3]

Figure 1 shows the illustration of the buoyancy principle.

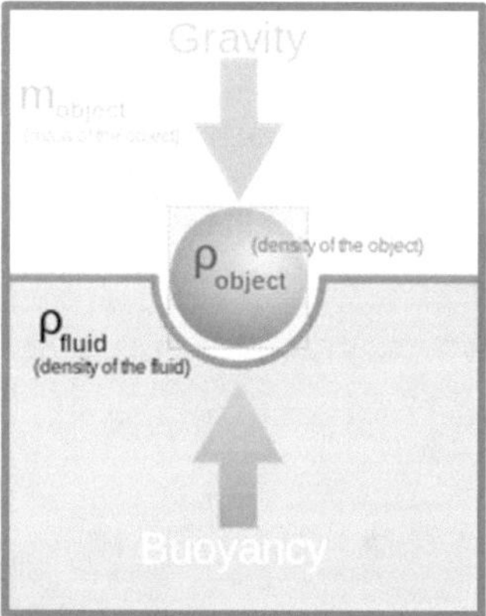

Figure 1. Bouyancy
Source: https://en.wikipedia.org/wiki/Buoyancy/

Archimedes Principle

Archimedes principle is the general principle of buoyancy. The *Archimedes principle* states that *"when an object is completely or partially immersed in a fluid, the fluid exerts an upward force on the object equal to the weight of the fluid displaced by the object"*.[4]

"The buoyant force acts at a point called the center of buoyancy, which is located at the center of gravity of the displaced fluid".[5]

"A completely submerged body displaces a volume of liquid equal to its own volume. Experience also tells us that when an object is submerged, it appears lighter in weight; the water buoys it up, pushes upward, partially supporting it somehow. The buoyancy principle asserts that an object

---

[3] Archimedes principle, https://www.britannica.com/science/Archimedes-principle

[4] Bouyancy: Floating and Sinking, Visual Physics, School of Physics, University of Sydney, Australia, *src.gov.jm/wp-content/uploads/2012/12/buoyancy.pdf*

[5] Donal F. Elger, et. Al, Engineering Fluid Mechanics, 10th Edition, Wiley, USA, 2013, p 87

immersed in a fluid will be lighter by an amount equal to the weight of the fluid it displaces. The upward force exerted by the fluid is known as the buoyant force."[6]

"If the weight of the water displaced is less than the weight of the object, the object will sink. Otherwise the object will float, with the weight of the water displaced equal to the weight of the object."[7]

As shown in Figure 2, the unloaded ship in "a" floats higher in the water than a loaded ship in "b".

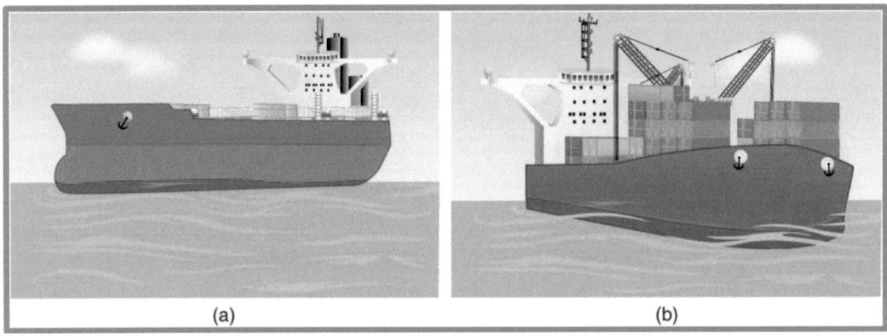

Figure 2. Illustration of Archimedes principle
Source: https://opentextbc.ca/physicstestbook2/chapter/archimedes-principle/

Buoyant Force Equation

Consider the illustration shown in Figure 3. Based on the illustration, the equation is derived:[8]

$$F_{up} = \gamma \, (V_b + V_a)$$

[6] Bouyancy: Floating and Sinking, Visual Physics, School of Physics, University of Sydney, Australia, *src.gov.jm/wp-content/uploads/2012/12/buoyancy.pdf*

[7] Archimedes Principle, https://physics.weber.edu/carroll/archimedes/principle.htm
[8] *Donal F. Elger, et. Al, Engineering Fluid Mechanics, 10th Edition, Wiley, USA, 2013, p86*

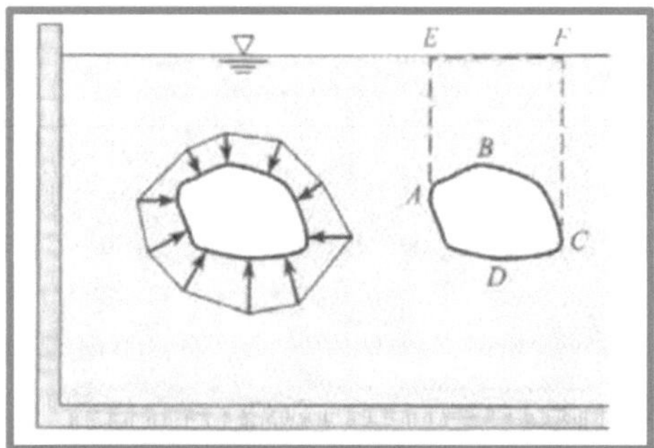

Figure 3. Buoyant Force
Source: Donal F. Elger, et. Al, *Engineering Fluid Mechanics*,
10th Edition, Wiley, USA, 2013

where:

F_{up} = upward force

γ = unit weight of the liquid

V_a = is the volume of liquid above the body

V_b = is the volume of the body

"The pressure acting on the top of surface of the body creates a downward force equal to the weight of the liquid above the body."[9]

$$F_{down} = \gamma \, (V_a)$$

where:
F_{down} = downward force

Therefore, subtracting the upward and downward force will result to a net force called the <u>buoyant force</u>.

$$F_B = F_{up} - F_{down} = [\gamma \, (V_b + V_a)] - \gamma \, (V_a)$$

$$\underline{\mathbf{F_B = \gamma \, V_b}}$$

where:
F_B = Bouyant force

Consider the illustration in Figure 4, wherein the body is partially submerged.

---

[9] Ibid

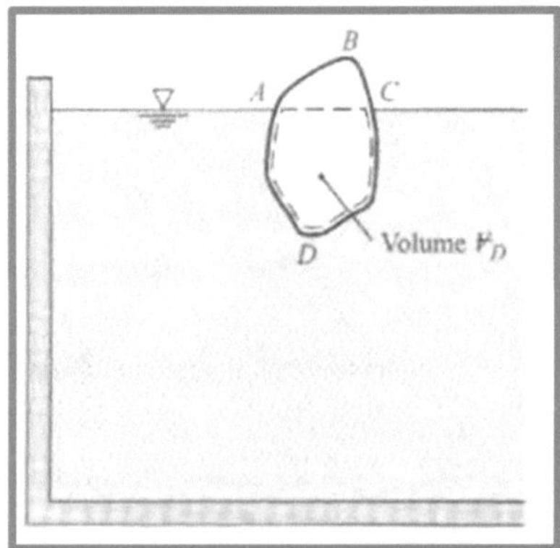

Figure 4. Partially submerged
Source: Donal F. Elger, et. Al, <u>Engineering Fluid Mechanics,</u>
10th Edition, Wiley, USA, 2013

"The marked portion of the object in Figure 4 has a volume of V_D. Pressure acts on curved surface ADC causing an upward force equal to the weight of liquid that would be needed to fill volume V_D. The buoyant force is given by:"[10]

$$\underline{\mathbf{F_B} = \mathbf{F_{up}} = \gamma\ \mathbf{V_D}}$$

Therefore, the buoyant force for a body that is partially submerged is determined by the V_D which is the displaced volume and is equal to the volume of the submerged portion and multiplied to the unit weight of the fluid.

"If the body is totally submerged, the displaced volume is the volume of the body."[11]

---

[10] *Ibid, p86*

[11] Ibid, p87

Problem Solving

Problem 1.

"A block of wood has a length = 2.5 m, width = 0.5 m and height = 0.4 m. The density of water is 1000 kg/m³. If the block is placed in the water, what is the buoyant force? Given the acceleration due to gravity as 10 m/s²."[12]

(Source: https://physics.gurumuda.net/buoyant-force-problems-and-solutions.htm)

Given:

Dimension of block:

      Length   =   2.5 m

      Width   =   0.5 m

      Height   =   0.4 m

Mass density of water (ρ) = 1000 kg/m³

Acceleration due to gravity (g) = 10

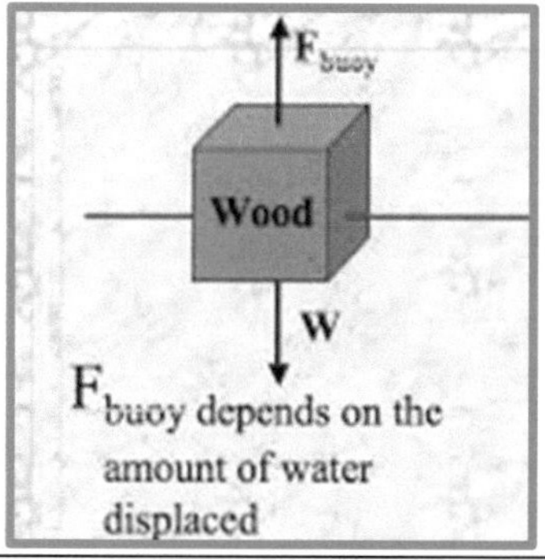

Figure 5:
Source:
https://web.suffieldacademy.org/~vkrasemann/Physics/buoyancy.htm

Solution:

Volume of the block (V)

V   = L x W x H =   2.5 x 0.5 x 0.4   = 0.50 m³

Buoyant force:

$F_B = \rho\, g\, V = \gamma\, V$

$F_B = (1000)(10)(0.5) = 5000$ Newton

$\underline{F_B = 5000 \text{ Newton or 5 kN}}$

---

[12] *https://physics.gurumuda.net/buoyant-force-problems-and-solutions.htm*

Problem 2:

"Weight of an object in air is 100 N. The object is placed in a liquid. The increase in volume of liquid is 1.5 m³. If the specific weight of the liquid is 10 N/m³, what is the weight of the object in liquid?"[13]

(Source: https://physics.gurumuda.net/buoyant-force-problems-and-solutions.htm)

Given:

Weight of object in air (W_{Air}) = 100 N

Increase in volume of liquid = Volume displaced ($V_{displaced}$) = 1.5 m³

Specific weight of the liquid (γ_L) = 10 N/m³

Solution :

Weight of object in liquid (W_{Liquid}) = $W_{Air} - F_B$

Weight of object in liquid = 100 N $- F_B$

$F_B = \gamma_L\, V_{displaced}$

F_B = buoyant force

γ_L = specific weight of liquid

$F_B = \gamma_L\, V_{displaced}$ = (10 N/m³)(1.5 m³) = 15 N

F_B = 15 N

W_{Liquid} (Weight of object in fluid)

W_{Liquid} = 100 N $-$ 15 N

W_{Liquid} = 85 N

---

[13] Ibid

<u>Problem 3:</u>

"A ship sailing in the sea enters a wide and deep river. The density of seawater is 1100 kg/m^3, the density of river water is 1000 kg/m^3. Determine the comparison of the volume of the object in seawater and in river water."[14]

(Source: https://physics.gurumuda.net/buoyant-force-problems-and-solutions.htm)

Given: Density of seawater $(\rho_1) = 1100 \text{ kg/m}^3$

      Density of river water $(\rho_2) = 1000 \text{ kg/m}^3$

<u>Solution :</u>

Since the ship is floating therefore the: Buoyant force (F_B) = Weight (W):

$$F_B = W$$

$$\rho_{water}\, g\, V_{displaced} = \rho_{object}\, g\, V_{object} ; \qquad \rho_{water}\, V_{displaced} = \rho_{object}\, V_{object}$$

$$\frac{V_{displaced}}{V_{object}} = \frac{\rho_{object}}{\rho_{water}}$$

$$V_{displaced} = \frac{\rho_{object}}{\rho_{water}} \times V_{object}$$

Volume of the object in seawater: Volume of the object in river water

$$V_{seawater} = V_{riverwater}$$

$$\frac{\rho\, object}{\rho\, seawater} \times V_{object} : \frac{\rho\, object}{\rho\, riverwater} \times V_{object}$$

$$\frac{1}{\rho\, seawater} = \frac{1}{\rho\, riverwater}$$

$$\frac{1}{1100} = \frac{1}{1000}$$

$$\frac{1}{11} = \frac{1}{10}$$

Therefore the ratio between the volume of the object in seawater and the volume of the object in river water is: **10 : 11**

---

[14] Ibid

<u>Problem 4:</u>

"A piece of equipment weighs 300 pounds on dry land. When the equipment is fully submerged in water the equipment weighs only 65 pounds. (1) Find the volume of the equipment, and (2) What is the specific gravity of the equipment?"[15]

(Source: http://www.peexamquestions.com/fluids-1/)

Solution #1:

$F_B = W_{air} - W_{submerged}$

$W_{air} = 300$ pounds

$W_{submerged} = 65$ pounds

$F_B = 300 - 65$

$F_B = 235$ pounds

$V_{displaced} = W / \rho$; where $\rho = 62.4$ lbs/ft^3

$$V displaced = \frac{235\ lbs}{62.4\ \frac{lbs}{ft^3}}$$

$V_{displaced} = 3.77$ ft^3

Since it is fully submerged, the displaced volume is equal to the volume of equipment

<u>**V_{equipment} = 3.77 ft³**</u>

Solution #2:

SG = Specific Gravity

$$SG = \frac{W object}{W water}$$

$$SG = \frac{300}{235}$$

<u>**SG = 1.28**</u>

Summary

In summary, the buoyant force is the net upward force on any object in any fluid and a result of subtracting the upward and downward Force.

Considering that the buoyant force is greater than the weight of the object, the object will rise to the surface and float. Otherwise, the object will sink.

If the buoyant force is equal to the weight of the object, the object will remain suspended at a certain depth.

Archimedes' principle states that the buoyant force on an object equals the weight of the fluid it displaces.

References:

Archimedes Principle,
https://physics.weber.edu/carroll/archimedes/principle.htm

Archimedes principle, https://www.britannica.com/science/Archimedes-principle

Buoyant force – problems and solutions,
https://physics.gurumuda.net/buoyant-force-problems-and-solutions.htm

Donal F. Elger, et. Al, Engineering Fluid Mechanics, 10th Edition, Wiley, USA, 2013

Bouyancy: Floating and Sinking, Visual Physics, School of Physics, University of Sydney, Australia, *src.gov.jm/wp-content/uploads/2012/12/buoyancy.pdf*

Donal F. Elger, et. Al, Engineering Fluid Mechanics, 10th Edition, Wiley, USA, 2013, p 87

Fluid mechanics, https://en.wikipedia.org/wiki/Fluid_mechanics

Fluids Problem (Buoyancy), http://www.peexamquestions.com/fluids-1/

http://site.iugaza.edu.ps/ymogheir/files/2010/02/Chapter1_Fluids_Properties_KA_YM_06Sep15.pptx

Y. Nakayama, Introduction to Fluid Mechanics, Butterworth Heinemann, Oxford, 2000, p13

Zerihun Alemayehu, Hydraulics I,
https://aaucivil.files.wordpress.com/2009/10/ch01.pdf